Examining
Wind Energy

Jordan Boyle

CLARA
HOUSE
BOOKS

**First published in 2013 by Clara House Books, an imprint of
The Oliver Press, Inc.**

Copyright © 2013 CBM LLC

Clara House Books
5707 West 36th Street
Minneapolis, MN 55416
USA

Produced by Red Line Editorial

The publisher would like to thank Ryan Light, Director of Renewable Energy,
Eastern Iowa College, for serving as a content consultant for this book.

Picture Credits
Fotolia, cover, 1; Shutterstock Images, 5, 10, 13, 14, 41; Muellek Josef/
Shutterstock Images, 8; Susan Montgomery/Shutterstock Images, 11; Peteri/
Shutterstock Images, 17; George Koultouridis/Shutterstock Images, 18; Pedro
Salaverría/Shutterstock Images, 21; David Gaylor/Shutterstock Images, 22–23;
Eugene Suslo/Shutterstock Images, 25; Red Line Editorial, 26, 34; Pavel Cheiko/
Shutterstock Images, 28; Thorsten Schier/Shutterstock Images, 30; Lucas
Payne/Shutterstock Images, 31; Amy Sancetta/AP Images, 33; Niek Goossen/
Shutterstock Images, 37; Julia Cumes/AP Images, 39; Maridav/Shutterstock
Images, 45

Library of Congress Cataloging-in-Publication Data
Boyle, Jordan.
  Examining wind energy / Jordan Boyle.
     pages cm. -- (Examining energy)
  Audience: Grades 7 to 8.
  Includes bibliographical references and index.
  ISBN 978-1-934545-47-8 (alk. paper)
  1. Wind power--Juvenile literature. I. Title.
  TJ820.B695 2013
  621.31'2136--dc23
                                    2012035318

Printed in the United States of America
CGI012013

www.oliverpress.com

Contents

CHAPTER 1

Endless Wind

Turning on the lights can be expensive. Have you ever heard your family complain about the high cost of electricity? Today, most of the world's energy comes from non-renewable sources. These non-renewable sources can have negative effects on the environment, and they will also eventually run out.

Innovators and scientists are always looking for ways to improve our sources of energy. Alternative energy research focuses on balancing our energy consumption needs against the needs of our environment. Finding renewable sources of energy to create electricity might also lower your family's electric bill.

Today, more than 80 percent of U.S. power comes from fossil fuels, the most common non-renewable energy source. Fossil fuels include oil, coal, and natural gas. They are made from organic material buried underground for millions of years. Because they take so long to make, we can't make more once we've used up our current supply. Wind power does not burn fossil fuels, so it creates less pollution. This renewable, clean

Seeking alternatives to fossil fuels, many communities are turning to wind for energy.

energy source is becoming more and more attractive to those concerned about the environment.

Wind farms use turbines to turn the power of wind into electricity. Humans have used the wind to do important work for thousands of years. While the supply of wind is unlimited, the strength of wind varies. Energy harvested by turbines must be captured and transmitted through cables to power the cities where people live. Some people are concerned about the impact of wind farms on natural landscapes, wildlife, and human health. Like all energy sources, wind power has limitations. But most energy experts agree that wind has an important place in our energy future.

EXPLORING WIND ENERGY

In this book, your job is to learn about wind energy and its place in our energy future. When did humans begin using wind power? How can the breeze you feel at the park turn on lights or power your computer? Can wind power alone meet our energy needs? What are the drawbacks of wind energy?

Megan Cruz is writing an article on alternative energy sources for her school newspaper. First she is researching wind energy. She will travel around the world interviewing experts and visiting scientists in the field. Reading her journal will help you in your own research.

CHAPTER 2

A Walk in the Park

I begin my investigation at the city park. It's a warm, breezy day, and many families are out enjoying the sunshine. Sailboats stream across the lake, racing faster with each gust of wind. Children watch their kites soar high in the sky. I can see the power of the wind all around me. To get started on my wind investigation, I'm meeting with Ken and Jasmine Peterson. Jasmine is a meteorologist, and Ken works for a group studying how neighborhoods can use clean energy sources. I'm hoping that between the two of them, I can learn a little more about wind and how people use it.

As we watch the drifting sailboats, Ken tells me a little about the history of the wind energy industry. "Interest in wind energy has been driven largely by concerns about the use of fossil fuels. Burning fossil fuels can harm the environment, which is a major concern. But there are also concerns about the

Have you ever harnessed the wind's energy by flying a kite?

cost and supply of fossil fuels. Gasoline is made from crude oil. In the 1970s, supplies of crude oil in the United States reached new lows. The low supplies accompanied by increased fossil-fuel usage worldwide led to a global energy crisis. The rising cost of gasoline reminded drivers that fossil fuels are a limited resource.

The more we use, the less we have. The less we have, the more it costs. Right now fossil fuels are pretty inexpensive as energy sources go. But as our fossil fuel supplies dwindle, prices will go up."

Ken tells me that in recent years, wind energy has emerged as a major alternative to expensive fossil fuels.

I say, "Because wind is free, right?"

"Not exactly," Ken explains. "Building a wind turbine can be expensive. But the U.S. government offers tax credits to people who install wind turbines for their homes or businesses. Once a turbine is built, it is relatively inexpensive to operate."

Ken tells me that governments are exploring new ways to spur growth in wind energy. Investors pay for much of the research that goes into developing new and better kinds of turbines. Ken explains that the cost of building and installing turbines has gone down in the past decade.

GREENHOUSE GASES

Burning fossil fuels releases carbon dioxide into the atmosphere. Carbon dioxide is a greenhouse gas. Greenhouse gases work a lot like a greenhouse for plants. These gases trap the sun's heat in the earth's atmosphere, making the planet warmer. Greenhouse gases help keep the earth warm enough for life. But large amounts of these gases in the atmosphere raise the planet's temperature too high, potentially causing changes in the climate that might harm the environment.

Once wind turbines are built, they cost relatively little to operate.

Jasmine chimes in, "Wind is an attractive energy source because it is completely renewable—it will never run out."

"How exactly does the wind work?" I ask. It seems like an easy question, but I'm surprised by the answer. Jasmine explains that wind is actually a kind of solar energy, created by the sun heating the earth's atmosphere. When the sun heats the atmosphere, the hot air rises very quickly. Cooler air rushes in to take the place of the rising hot air, creating wind.

"But what causes gusts of wind and breezes?" I ask.

"Air is made of tiny, invisible molecules," Jasmine says. "You can feel the force of their weight against your skin as they move. Sometimes these molecules are pressed close together, or under high pressure. Other masses of air have molecules that are farther part, or under low pressure."

Jasmine tells me that air is always on the go, moving from areas of high pressure to low pressure. Currents of air swirl around our planet—sometimes in gentle breezes, sometimes in fierce gusts. Hills, mountains, valleys, and even the earth's rotation create the gusts and breezes. This movement of air is a kind of energy called kinetic energy. Kinetic energy is the energy something possesses because of its motion. A roller coaster speeding down a hill has kinetic energy.

"Days like today are great," Jasmine adds, "but as you've probably noticed, it's not always windy. This poses a problem for wind energy. We need to find a way to capture energy from wind and store it, even when the wind isn't blowing as hard as it is today."

As I listen, I feel the breeze as a weight against my skin. Who knew we were all under so much pressure! But I wonder, how can kinetic energy be used to do work?

Wind energy is easy to capture on breezy days. But what happens on a day when the air is still?

CHAPTER 3

A Windy History

Today I've traveled all the way to the shores of the Red Sea to find out how humans can harness the power of the wind. Between Africa and Asia, the Red Sea is an important stop on a tour of the history of wind power. I'm meeting with archaeologist Humphrey Daniels to find out why.

It is a busy day for Humphrey and his team. He is directing workers and teams of archaeologists and scientists as they build a ship modeled after an ancient Egyptian sailing ship. He explains that humans began using the power of wind as early as 2800 BCE. The Egyptians hoisted sails on ships that carried them along the Nile River or across the Red Sea. They could go as far as 2,000 miles (3,200 km) in one trip.

I remember the sailboats gliding across the lake at the park. Those modern boats also moved through the water by catching the breeze in their sails. "How do we know Egyptians used this method so long ago?" I ask Humphrey.

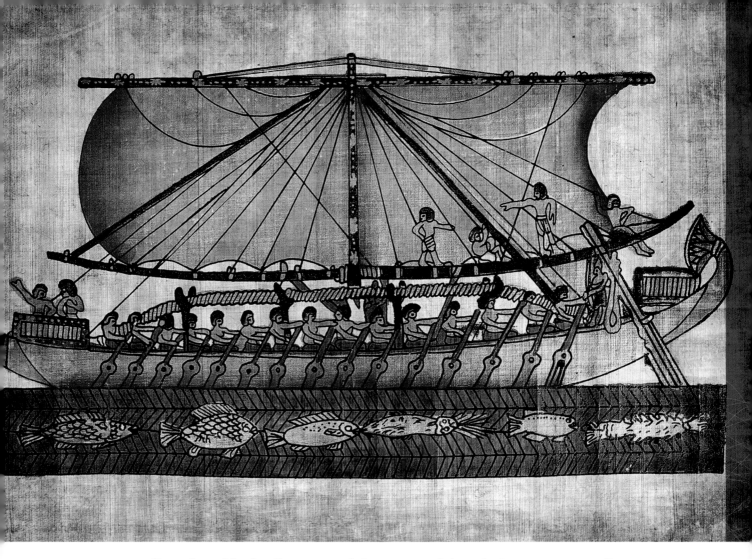

Egyptian ships' sails captured the power of the wind to transport sailors across bodies of water.

"We've found ancient drawings and sculptures of sailing vessels in Egyptian temples," Humphrey explains. "A culture's art often reflects what is going on in the everyday life of its citizens. So we know ancient Egyptians used sailboats. Before sailing, people used oars to row boats over long distances. This no doubt made for some exhausting journeys! Sails allow boaters to let the wind do a whole lot of the work. I'm using the drawings we've discovered to try to recreate an ancient Egyptian boat that will hoist its sails on the Red Sea."

The ancient Egyptians sound pretty smart. But Egyptians weren't the only people who sailed into wind energy history. Ancient Persians who lived in modern-day Iran used windmills to grind grain. Humphrey describes how Phoenicians in the Middle East and Europeans also used sailing ships to travel the globe before steam-powered ships came along in the 1800s.

A strong wind comes off the shore, and I imagine these early sailing ships using the power of wind to travel the world and make important discoveries. I wish Humphrey good luck on his mission to recreate the earliest example of humans using the power of the wind. I'm off to one more stop on a tour of wind energy history.

EARLY SAILING VESSELS

Throughout history, sailors have used different kinds of ships to harness the power of wind in different ways. In the 1400s, Portuguese ships sported lateen-rigged sails. These were triangular sails that ran parallel to the sides of the boat. Lateen-rigged sails caught wind gusts to travel fast, and they were easy to steer along a coast. Oceangoing ships that needed to travel long distances often used square-rigged sails. These were square sails that ran perpendicular to the edges of the boat. Square-rigged sails used steady wind to travel far.

People still use wind-powered sailboats to travel Egypt's Nile River.

CHAPTER 4

Inside a Windmill

Today I am traveling by ferry along a set of canals outside Amsterdam, the capital city of the Netherlands. It's a beautiful day in the countryside. I'm with a group of tourists from all over the world. We excitedly await our destination, and it appears around the next bend.

Set against a blue sky, I see the wooden blades of a windmill twirl just as they did 200 years ago. I've come to this windmill for a tour of one of the most famous examples of wind power at work.

Our tour guide, Anna Van Dijk, greets us as we disembark. She explains that ancient cultures didn't just use wind power for sailing. Since as early as 600 CE, humans have used wind power to help with tasks such as pumping water and grinding grain. "The Netherlands is very close to sea level," Anna explains. She points to the windmill. "Much of our country used to be under

The windmills of Kinderdijk-Elshout, in the Netherlands, drained water from areas that could be turned into farmland.

water. In the 1400s, the Dutch used windmills like this one to pump water away from lowlands. They took land that was once under the ocean and turned it into valuable farmland. Let's take a closer look at the mill."

The power of the wind turns giant gears inside the windmills to pump water.

A ghostly hum and whine greets us as we near the structure. Anna explains that this is the sound of the turning blades converting wind into kinetic energy. Then Anna leads our group inside the mill.

The windmill we are in is like many of the old windmills in Europe. It consists of four rotating blades set on top of a wooden tower. High inside the windmill, I see a horizontal shaft that is turned by the blades we saw outside. "This is called the windshaft," Anna explains. She points out how the windshaft turns a creaking set of gears. The gears rotate a larger upright shaft that reaches down below the floor. Gears at the bottom of the tall shaft turn a giant screw we can see through a window at our feet.

"But how does all this work to pump water?" I ask.

"The screw through that window draws water from the lowland into a storage basin, draining land that can be used for farming," Anna explains. The gleaming screw rotates at our feet, dripping water as it turns. "Here the kinetic wind energy outside is turned into mechanical energy, the energy used to do work."

It's amazing to think that hundreds of years ago people could have created such a complicated structure that still works today. But I still have many questions. How did these early windmills that turned wind power into mechanical energy evolve into the wind turbines of today? How is a wind turbine different from a windmill? Next stop: a modern wind farm.

AMERICAN WINDMILLS

Old windmills weren't exclusive to Europe. American farmers used windmills for pumping water on the Great Plains in the 1800s. Tens of thousands of windmills dotted the landscape of rolling hills and farmland. These windmills were constructed with sturdy metal blades instead of wood to hold up under the intense power of the prairie winds. Farmers used these windmills to pump water for livestock and crops. Many of these old mills are still in use today.

CHAPTER 5

Today's Turbine

To find out how a modern wind turbine works, I've come back home to the United States for a visit with wind technician Cory Nasser on the plains of west Texas. As I get closer, I can see the wind farm from a distance. There are dozens of three-bladed turbines turning in the wind. Each turbine must be at least 80 feet (24 m) tall. When I step out of the car, the wind almost blows my cowboy hat off.

I know that wind can be noisy. But I notice another sound as I get closer to the turbines. I hear a kind of hum or swooshing—it sounds a little like a washing machine. Cory shouts to me that the blades swooping through the air create this sound, and he'll tell me more in a minute. As we get closer to the first turbine, which is named T-13, it gets much quieter. Cory explains that this is because the sound of the turbine is broadcast outward.

"That's one of the reasons we try to build wind farms far away from cities and houses," he says. "Still, it's not

Wind farms can generate a high volume of wind energy.

always possible to put a wind farm in a completely isolated location. The noise can really bother our neighbors. Studies have shown that the noise of a wind farm does not pose a risk to human health. Even if the turbines don't cause physical

damage, people may not want to live by them. However," he adds, "turbines keep getting better. The newer ones are much quieter than older ones. Wind farms can have a few turbines or hundreds. Big farms like this will always take up a lot of space.

Wind farms can take up large tracts of land.

The land around the turbines doesn't have to be wasted, though. Animals can graze on it, or crops grown."

I ask Cory how this sleek turbine compares with the Dutch windmill I toured. "It's the same basic idea," Cory says. "But there are some major differences. The windmill you saw converted the energy of the wind into mechanical energy to run that water pump. These turbines turn the power of the wind into electricity. Basically, it's the opposite of a fan you plug into the wall. Instead of using electricity to create wind, a turbine uses the wind to make electricity."

To show me how the turbine creates electricity, he points out the main parts of the structure we can see from the ground. "There is the giant tower, the rotor with its blades, and the nacelle. The nacelle is the horizontal casing on top of the tower," Cory explains. "The nacelle is where the magic happens!" He hands me a hardhat and harness and opens the door to the tower.

"You might think strong winds are a good thing for a turbine," Cory says. "But winds that are too strong can damage the blades. A turbine like this one has computer controls that move the pitch, or position, of the blades to protect them in strong winds. Sensors at the top of the turbine on a device

WIND FARMS AT SEA

One challenge for wind farms is how much space they take up on land. But some of the most amazing modern wind farms rise directly out of the sea. Ocean winds are often stronger and faster than onshore winds. This is because there are no barriers such as trees or mountains to break the wind. Offshore wind farms feature groups of turbines that are often driven right into the sea floor. The Walney offshore wind farm in the Irish Sea is one of the largest offshore wind farms in the world. It has more than 100 turbines sending electricity through cables to a station on land more than nine miles (14 km) away.

Offshore wind farms are built to capture the winds that whip up across the open ocean.

called an anemometer measure the wind's speed and direction. This turbine is also designed so its blades rotate in the opposite direction from how the wind is blowing."

Cory helps me clip my harness to safety cables as we climb a seemingly endless set of stairs up the tower. We stop to catch our breath at several decks along the way. At the final deck, Cory opens a steel trapdoor to a tremendous hum of machinery. We are inside the nacelle.

Next Cory opens a hatch in the nacelle, which we wiggle through to stand directly on top of the giant turbine. I marvel at the enormous blades spinning and the beautiful scenery beyond. "Some people think wind farms are an eyesore," Cory says. "And there's no doubt they change the look of the natural landscape. But I think they can be beautiful, don't you?"

I've learned a lot from Cory and T-13. But I have a couple more stops to make before I truly understand wind energy.

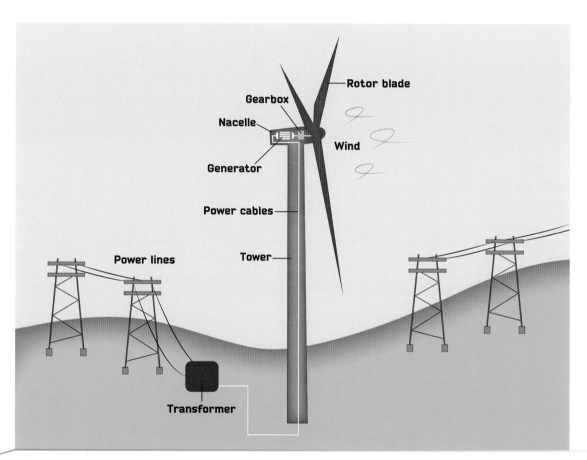

A WORKING WIND TURBINE

At the front of the nacelle, a low-speed shaft is turned by the rotation of the blades just outside. This shaft runs into the gearbox. Inside, two gears work to speed up the turning of the shaft. The shaft runs into the giant generator. The generator creates an electric current. Then a wire transports the electricity from the turbine to the transformer.

CHAPTER 6

Catching the Wind

Katrina Strand is a program manager at a renewable energy research group. She keeps up on the latest developments in storing and using wind power. Today she is showing me around an abandoned mine pit on Minnesota's Iron Range. The former mine is filled with water. It looks like a lake. There are few trees or plants and no houses or buildings as far as the eye can see. Every now and then a gentle breeze ruffles my hair. "What does an abandoned mine pit have to do with wind energy?" I ask Katrina.

"Good question," she answers. "It begins with the variability of wind power." She explains that wind is intermittent, meaning it varies in intensity. The breeze we feel on this day, for example, comes and goes. It may be completely calm a few miles from here.

The wind is not always blowing. One of the challenges of wind energy is finding a way to use it on very calm days.

"Unfortunately," Katrina says, "you can't put wind in a box. One of the challenges of using wind power is getting a steady supply of electricity to homes."

I've been wondering about this since I visited the wind farm in Texas. "How does electricity from a wind farm get to a home?" I ask.

Katrina explains that electricity from the turbines I saw in Texas moves through something called the electrical grid. First, the electricity made by a wind farm travels through a series of wires to a substation. The substation has a group of transformers. The transformers raise the voltage, or the strength of the electric current. The higher voltage makes it possible for the current to travel over long distances. The current then goes through high-voltage power lines to another group of transformers. These lower the voltage back down.

"From the transformers, the current travels though those electrical wires you can see near your house or beside a road," Katrina says. "A transformer somewhere near your house further lowers the voltage to a level you can use inside. Underground wires carry the current to a box called a circuit breaker or fuse box. The circuit breaker can cut the current in an emergency. From there, wires carry the power to those handy outlets you use every day."

THE ELECTRICAL GRID

What is the electrical grid? You might be able to see it right now if you look out your window. The electrical grid is a network of wires connecting towns, cities, and rural areas all across the United States and every other developed country. You're probably so used to seeing these wires on their tall poles that you don't even notice them anymore. With the electrical grid all around us, it's easy to forget what amazing work it does.

Katrina says the system works pretty well, but transporting the electric current can be really expensive. Since most wind farms are located away from residential areas, miles and miles of wires are needed to move the power from the wind farm to a home.

I still don't see what this has to do with abandoned mine pits. By now, I've taken a rest on a large rock, but Katrina is

The electrical grid brings electricity from its source to where it is used.

just getting warmed up. "You see, if wind were always blowing at the same speed in the same place, it could supply a steady stream of electricity to your house. But wind is not that steady. And you need different amounts of electricity at different times.

"To make a long story short," she says, "we hope to use mine pits such as this one to store electricity generated by wind. Wind often blows strongest at night. We would use this extra energy to pump water from the pit uphill to a storage basin." I immediately think of the windmill I visited in Holland. It did the very same thing. Katrina goes on to explain that during the day, when customers use more energy, the water would be released. The rush of water would create energy of its own, ready to meet the demand of customers.

And this isn't the only way to use water to store wind energy. Katrina tells me that some companies are storing wind energy in underwater balloons. There, the energy is out of sight and ready for use when demand goes up.

Lakes left over from former mines, such as this one in Virginia, Minnesota, might one day store wind energy.

"So that's it," I say. "The answer to storing wind power is water!"

Katrina explains that these are new ideas that still need testing. Another method of storing wind energy involves no water at all—battery backup. Batteries are a way to store electrical power. Just think of when the power goes out in your home. An alarm clock with a battery will still work. An alarm clock plugged into the wall won't. A battery that could store enough wind power to even out the supply to homes would need to be much, much bigger than an alarm clock battery. Some companies are testing giant batteries for this purpose.

Katrina has given me a lot to think about. Every day, scientists are thinking up more ways to store wind energy!

CHAPTER 7

All Kinds of Turbines

In my travels, I've stood inside an old Dutch windmill. I've climbed on top of a sleek, modern turbine. But I am about to see a different kind of turbine. Today I'm sitting in the office of Don Carrol. Don is a senior researcher at the Institute for Wind Power. Don's office is outside of Chicago, Illinois. I look out the window, and there's not a turbine in sight. Don explains that the large turbines on wind farms are just one way to make electricity from wind. Smaller turbines, and even some structures I wouldn't recognize as turbines, are helping to meet people's energy needs all over the world.

"Here," he says, "try this out." He hands me a small wind turbine—so small I can hold it in my hand. I blow on the blades as if it were a pinwheel. To my surprise, a green bulb on the back of this mini-turbine lights up. It's working! Using only the air from my lungs, I've just generated enough electricity to light a bulb!

"Around the world, there are also thousands of small turbines making electricity," Don continues. "These are usually used for homes and small businesses. Very small turbines called microturbines do even smaller jobs. For example, nomadic groups in Mongolia use microturbines to create the energy they need to boil water. These turbines are small enough to be carried on horseback. You might spot a microturbine at the local park. They are sometimes used on sailboats to charge

In 2012, the Cleveland Indians became the first Major League Baseball team to install a wind turbine. Wind catches the microturbines in the corkscrew to help power the stadium.

the boats' batteries. From sailors on the open sea to villagers in remote areas, small turbines do work where it's not possible to connect to the electrical grid."

"How do these small turbines compare to the large ones I visited in Texas?" I ask. "How powerful are they, really?" Don explains that the power ratings of turbines depend on how big their blades are. The more area covered by the blades, the

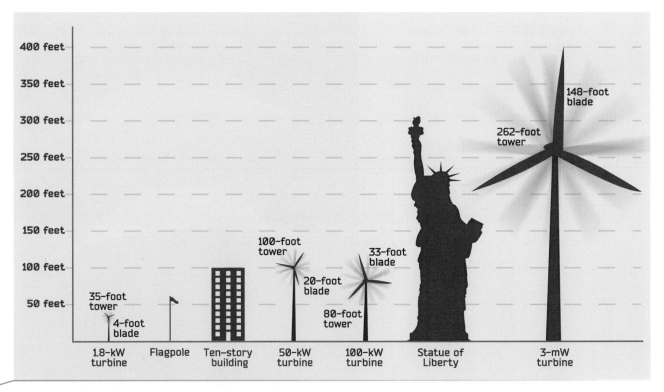

HOW TALL IS A TURBINE?

Some wind turbines are small enough to hold in your hand. Others tower above the Statue of Liberty. A ten-kilowatt turbine could produce all the electricity you need in your home. The largest turbine ever built has blades that are each longer than nine buses. This giant turbine produces enough energy to power almost 2,000 homes.

more electricity a turbine can produce. So, large turbines are usually more powerful than small ones.

"But that's not all," Don continues. "Turbines come in all sizes, and they also come in all shapes." Don explains that all the turbines I have seen so far are called horizontal-axis turbines. On horizontal-axis turbines, the shaft turned by the blades is horizontal to the ground. However, some turbines are attached to a vertical shaft. Other turbines have no blades at all! One kind of turbine flies in the sky like a spinning blimp.

Engineers are constantly trying to improve the design of wind turbines. They work to make turbines more effective and cheaper. Giant turbines, microturbines, turbines without blades—who knew there were so many ways to capture the power of wind!

TURBINE POWER

Turbine power is given in a number known as rated power. The rated power of a turbine is the amount of electricity it can produce. Electricity is measured in units called watts. One thousand watts equals one kilowatt. One thousand kilowatts equals one megawatt.

CHAPTER 8

Wind and Wildlife

Today I'm in Nebraska meeting with Mila Thompson, a field researcher for an environmental group that studies the impact of wind turbines on wildlife. Her area of expertise is raptors, which are birds of prey. Today her team is studying migrating birds outside Omaha, Nebraska.

Mila tells me a big concern people have about wind turbines is the threat they pose to birds and bats flying in the area. Raptors, such as hawks, are especially at risk. When a hawk is focused on hunting mice in a field, it may not see a turbine.

"Wind energy can be a good thing," she explains. "While turbines might sometimes leak oil or other chemicals into the ground, wind is overall a much cleaner source of power than fossil fuels. Any energy source that does not pollute is a good thing for wildlife."

No energy source is perfect, but wind energy has the potential to provide a lot of energy in the near future.

She adds, "However, scientists believe approximately 200,000 birds per year are killed by wind turbines. While this number is big, it is much smaller than the number killed each year by power lines, cars, and other man-made objects. The bad news is that turbines seem to affect large birds of prey disproportionally, such as eagles and hawks. Building wind farms in the right places—away from migration routes, for example—could help cut down on these bird fatalities. My team's job is to study proposed wind farm sites to see if they pose a danger to birds."

Mila tells me that a well-chosen site for a wind farm would have little effect on the surrounding wildlife. I remember what Cory told me, that livestock such as cattle can even graze beneath wind turbines! But controversy still surrounds the building of new wind farms. The Cape Wind project off the

WIND POWER BY COUNTRY

The amount of wind energy that can be produced in a country is its installed wind capacity. China, the United States, and Germany have the highest installed wind capacities. However, European countries lead the way in producing wind power. Almost 20 percent of the electricity generated in Denmark comes from wind. Portugal is close behind at 18 percent.

island of Nantucket in Massachusetts is one example. In 2010, government officials approved the construction of what would be the first U.S. offshore wind farm.

The plan to build this wind farm was controversial. Local residents worried that the farm would change the scenery, negatively affecting tourism and property values. People also worried about the effect it would have on the environment. Those who fished for a living worried about its effect on the fish in the area, although some studies have shown that fish actually thrive near offshore turbines. The wind farm would potentially meet 75 percent of Nantucket's energy needs, but because wind farms are so expensive to build, residents worried electricity costs could go up. Still, several major environmental groups spoke out in favor of the wind farm.

Wow, 75 percent sounds like a lot of energy. I ask, "How much energy do we need? Could we get all of our electricity from wind one day?"

Wind energy can be controversial. Many people had strong feelings about the wind farm proposed off the coast of Nantucket.

"Maybe," Mila explains, "Just 10 percent of the total wind energy on Earth is much greater than the energy consumed worldwide." She tells me the U.S. Department of Energy is studying how the United States might be able to get 20 percent of its energy from wind by the year 2030.

From what Mila has told me, wind energy may have its drawbacks. But with careful planning and research, new wind farms can be built in areas where danger to wildlife can be minimized. With scientists and engineers thinking of new ways to capture and store wind energy, there's no telling what wind energy could do in the future.

CHAPTER 9

Your Turn

You have had a chance to follow Megan on her trip around the world to study wind power. Now it's time to think about what you have learned. Wind power is a good alternative energy source because it is renewable. But the wind doesn't always blow at the same speed or intensity. Wind energy must be stored and sent through the electrical grid for people to use. Wind power creates relatively little pollution and has the potential to provide more energy than we could ever use. But some people worry about the dangers large turbines pose to the environment and wildlife, especially birds. Some people think wind farms are ugly. This might affect tourism and property values. But with innovators and experts constantly working to improve wind power, the technology keeps getting better. Every year, more of the world's electricity is coming from wind.

YOU DECIDE

1. Do you think the pros of wind energy outweigh the cons? Why or why not?

2. Do you think Nantucket Island should move forward with the Cape Wind offshore wind farm? What are the reasons for your answer?

3. What can you do to cut down on your energy use? Think about technology, such as solar panels, and also ways to change your behavior, such as walking instead of driving.

4. Wind farms can be dangerous for wildlife. Is the cheaper, cleaner power we can get from wind farms worth the risk? Why or why not?

5. How could wind energy be used in your home or school? Where would you consider building a wind turbine?

How big of a role do you think wind energy should play in our energy production?

GLOSSARY

alternative energy: A type of energy that is from a renewable source that is not in danger of running out, such as the sun, wind, or water.

atmosphere: A layer of air and gases surrounding the earth.

current: In science, the flow of electrical charge through a substance.

electrical grid: A network of cables that delivers electricity.

fossil fuel: An energy source that developed from the remains of animals and plants.

generator: A machine that converts mechanical energy into electrical energy.

kinetic energy: The energy of motion.

mechanical energy: A type of energy that does work.

molecule: The smallest unit of a substance that still possesses the properties of that substance.

organic: Material from living plants or animals.

renewable: When something can be replaced by natural cycles in nature or the environment.

tax credit: An amount of money subtracted from the total fee collected by the government.

turbine: A machine that converts wind energy into electrical energy.

watt: A unit of power for measuring electricity.

windmill: A machine that does work, such as pumping, using the energy of the wind.

EXPLORE FURTHER

Track the Wind

Use the Internet or your library to learn more about the jet stream, air pressure, and wind speeds. Then, watch the weather forecast on your local news every night for one week. Keep a journal recording every time the meteorologist uses these terms and what the weather is like each day. How might these forecasts affect windmills in your area?

Visit a Wind Farm

Do you live near a wind farm? Wind turbines may be closer than you think. Use the Internet and your local library to research nearby turbines and wind farms. Then see if the turbines are open to the public. You might be able to sign up for a tour. If not, it's fun to just go take pictures and listen to the sound of wind at work.

Shrink Your Carbon Footprint

The amount of carbon dioxide you produce is sometimes called your carbon footprint. Use an online carbon-footprint calculator to estimate how much carbon dioxide your household produces in a year. Examine your results. Where can you reduce emissions? Can you hang laundry in the sun instead of running the clothes dryer? Can you replace outdoor lighting with

Visiting a nearby wind farm can be a great way to learn more about alternative energy.

solar-powered lamps? What about growing your own food to reduce trips to the grocery store? What are some other things you could do to reduce emissions?

SELECTED BIBLIOGRAPHY

Chiras, Daniel. *Power from the Wind*. Gabriola Island, BC: New
    Society Publishers, 2009.

Gipe, Paul. *Wind Energy Basics*. White River Junction, VT:
    Chelsea Green, 2009.

Hsu, Tiffany. "Wind Turbines Growing Taller and More
    Powerful." *Los Angeles Times*, July 24, 2011. Web. Accessed
    September 11, 2012.

Jones, Mike. *Wind Energy*. Waco, TX: TSTC Publishing, 2010.

FURTHER INFORMATION

Books

Goodman, Polly. *Understanding Wind Power*. New York: Gareth Stevens Publishing, 2011.

Ollhoff, Jim. *Wind and Water*. Edina, Minnesota: ABDO Publishing, 2010.

Morris, Neil. *Wind Power*. Mankato, Minnesota: Smart Apple Media, 2010.

Websites

http://www.youtube.com/watch?v=ma8pCwPS-Uk
Go inside a 200-year-old Dutch windmill to see and hear how it works.

http://kidsahead.com/subjects/2-wind-energy/activities
This website lists experiments you can do at home to better understand wind energy.

http://www.pwrc.usgs.gov/resshow/windpower/
Visit the Patuxent Wildlife Research Center website to learn more about the impact of wind farms on wildlife.

INDEX